# 2012:

# The One Heart Transformation

# Four Discourses

by

Terrence Deaton Bell

ISBN: 0615456146
ISBN-13: 978-0615456140

# DEDICATION

To Necia Lynn Nicholas who models the One Heart life everyday

# CONTENTS

## Cover

The front and back covers of this text are partials of two murals by Danish artist Jørn Bie. The full murals hang in Antigua, Guatemala's restaurant, Dona Louisa. Jørn Bie spent 17 years doing sketches and murals of Guatemalan indigenous life. His drawings and these murals are the finest metaphorical representations of indigenous life, spirit, and promise.

# Introduction

These four discourses on 2012 are just that, formal conversations. They are related to the Mayan Calendar's end of a very important time group. There have been many books, web sites, and organizations formed around the Mayan Calendar and the Mayans themselves. The contemporary groups of indigenous people of Central America who are generally considered to be decedents of the ancient Mayans are the constant study of individuals and organizations ranging from serious academics to interloping practitioners of Mayan life and Mayan spirituality. My wife, the author of this work's preface, is one of those serious academics. I have to admit that it is, in fact, her work in Mayan culture which gave me entrance into the sometimes very private lives and ceremonies of Mayan Shamans and people. This is not to separate, as so many do, the indigenous people of Guatemala from the rest of Guatemalan society. We value them all and recognize their continuing effort to form a nation state together.

This does not put aside the difficult history of Guatemala with respect to a 35 year civil war and past and on-going prejudices between indigenous and Ladino Guatemalan society. Nevertheless, these discourses are not about that important and on-going struggle. In fact, this work is not about many of the rather usual views or studies about this important Central American country. My purpose in these discourses is not to overwrite or explain some of the interesting complexities of the Mayan. Most notably I will not wholly explain Shamanism, how the

Mayan Calendar works, how to read Mayan iconographs, the Cofradia, the intricacies of Maximon or Mam, or engage in wild guesses about 2012.

When conversing with Mayan Shamans and often Mayan people in general, I would hear some conflicting views on a year of great interest to the Mayan Calendar, 2012. Also, early on in my stays in Guatemala I saw many books and pamphlets about the Mayan Calendar's 2012 year as delivering some fantastic, even apocalyptic, events. Because I had the opportunity to be with and close to so many Mayan people, I became increasingly interested in their various views of 2012. In the end I discovered that the 2012 date could be viewed with some interesting complexities. Yet, I would expect no less from a people who see life in deep mathematical terms. I most often heard 2012 explained in terms of the number 13, which has multilayers of importance. In short, for the Mayan, 13 is the number of completion (much like Christianity's number 7). Their calendar count goes from 0 to 13 and after 13 the count resets to 0. Also, 1 to 13 are the coefficients they have for each day sign or "month" on the calendar. Therefore, the calendar count only goes up to 13 and therefore resets. Simply, resetting or not, this is an important end time count. I think we all have an interest in end times. Whether this is a matter of an end that signifies the beginning of something new, or a hanging on to the importance of religious apocalyptic ends, or knowing that our death is an end we hope is not "thee end"; when we hear of an end from what we think are some mysterious peoples, we may think we have stumbled upon some fantastic hope.

It is in part these fantasies I rather address in the discourses. But the discourses are not about the many interesting guesses about the year 2012. I honestly don't embrace any of them eventhough I have enjoyed the adventure of reading all of them. As you will see in the discourses, I have dismissed them as rubbish. My time with Guatemalans gave me a very different interest in the Mayan Calendar.

In the end I am surprised at how often I leaned on Miguel Leon-Portilla's *Time and Reality in the Thought of the Maya* for dealing with the discourses' issues and finality. Perhaps this is because my mind if not my heart works in the academic world. I believe anthropologists, linguists, epigraphers, and culturists are doing the important work in Guatemala. This belief is no doubt rooted in my hanging about with these academics whether in their print or person. Nevertheless, these discourses are from my contact with Guatemalans, indigenous and Ladino. My own wife's PhD work as a Mayan culturist should not go unrecognized as well as Dr. Judie Maxwell of Tulane University, who is doing the best work for indigenous people in all of Guatemala. The influence of Michael Coe, Linda Shele, and Miguel Leon-Portilla's wonderful writings as anthropologists and epigraphers is undeniable.

Still, these discourses are not in the realm of these academy writers and workers in Guatemala. I was moved to these discourses by my loving contact with Guatemalans, both Ladino and Indigenous. I was welcomed to their homes, their churches, and their ceremonies—both public and private. What I observed are a people like all of us--people who care about their nation, their families, their jobs, and their personal importance. What is considered "mysterious" about Guatemalans is their indigenous Mayan population who are a majority of the Guatemalan people. The Mayan history is one of antiquity and mystery. The complexities of their calendar system are a great part of this somewhat mystical people.

My purpose in these discourses is not to overwrite or explain some of the aforementioned interesting complexities of the Mayan. I simply want to offer a genuine gesture from the only people who are original to our cherished American lands. Given this, I will acknowledge some of these wonderful people who made these discourses possible at the end of the resources page. It is through them that the discourses take up the true, constant theme of all these people: a One Heart transformation.

# 2012: THE ONE HEART TRANSFORMATION

# Preface

*Elizabeth R. Bell*
*The Ohio State University*

There are changes occurring in Guatemala. After a decades-long civil war that followed three tumultuous centuries of political and social upheaval, the Guatemalan people are in the process of searching for balance in their nation state. This task is not a simple one, due in part to the dual cultural influences of both the Mayan peoples and the Ladino – or *mestizo*, of mixed Spanish descent – population. Although census data is conflicting, the Mayan population is generally considered to comprise around 60 percent of the overall population. Nevertheless, although they are the dominant group in number, the Mayan population has historically lacked political and economic power in the country. While the Guatemalan Constitution as well as the Rights of Indigenous Peoples Act (1995) aim to guarantee the rights of the indigenous populations of Guatemala, these documents also limit the Mayans' belief systems and ways of life to the realm of folkloric banter. The Guatemalan nation-state has no reservations about using the Mayan "traditions" as its main tourism marketing strategy, but this same supposed reverence for these customs lacks any real economic or political benefit to the people who practice them.

In recent years, the international forum has entered into the dialogue with the representation of Mayan customs. This is mostly due to the approaching date of December 21st, 2012 on the Mayan sacred calendar, which signifies the end of 13 approximately 400 year cycles. Prophetic claims abound, predicting anything from death and destruction, to a shift of consciousness, to a time period when the Mayans will take back what was stolen by the *invasores* – the invaders, the Spaniards, who arrived right around the end of the last cycle –, to absolutely nothing. Mayan shamans do not agree among themselves what this date may hold, and international authors certainly do not, either. The 2012 phenomenon is primarily about what all prophecies purport: hope for change. That change may be positive or it may be negative, but in any case the expectation is that something will be forever different. This is an attractive concept to people in general because of our desire for sensationalist musings combined with our constant search for something meaningful. This is why so many different groups of people, from the local all the way to the global, have been attracted to the idea inherent in 2012. The endemically Mayan concept of 2012 has moved from the original meaning it may have had into a field of multivalent representation in which many voices offer their opinions.

This book does not make a prophetic claim. In fact, Terrence Bell asserts that a central step in understanding the 2012 prophecies as he does is to view them not as prophecies at all, but rather as prefigurations. This book does not claim to explain what the Mayans really mean by 2012. Instead, he uses what he has observed during his time in Guatemala with the Maya to give his readers a didactic tool, much in the style of an oration or oral lesson. Here it is fitting to reveal that, originally, Terrence did not call these four chapters "four discourses." They were called sermons. This original conception of the work is quite revealing of the nature of this text, as what he has written hints at the sermon genre. As such, it is also fitting to note that the style of this book is really best suited to oral transmission rather than written. To explain further what he sets out to do here, we will understand sermons as an oral genre which transmits a personal yet authoritative interpretation on a concept of a religious nature, with the goal of teaching something to its listeners. This book provides purely that: an explanation of how the 2012 concept can be used in a personal narrative for the individual. He does not claim to be Mayan, to speak for the Mayans, or to know better than the Mayans. Instead, he explains in these pages his personal, and one could argue authoritative due to his time spent in Guatemala as well as his degrees in Theology and Rhetoric, interpretation of a religious or spiritual idea. Neither does he ask the reader to assume a great responsibility or to become Mayan. He simply asks them to breathe.

# Discourse One:

# The Meaning of Mayaness

These discourses are related to a question to which you already know the answer. The question is, "What is the most important thing in the world?" There is a right answer. The answer is easy, and if you do not agree with my answer, I hope somewhere along the way or by the end of Discourse Four you will not only agree with the answer but also become an active member of the 2012 transformation. Discourses typically aim to inform us. Here, I want to present information on the transformative stage that is 2012. In fact all the writings, movies, and talk about 2012 are centered around transformation, and all those efforts are false. They are false because they depend on illusions about 2012 rather than Mayan authenticity. In these discourses, we will look at Mayan reality as the truth about 2012. I have fashioned these as discourses rather than polemics because 2012 is an event often discussed in religious or apocalyptic terms rather than in Mayan authentic belief systems. 2012 is about a spiritual reality of action not about illusionary manifestations. Therefore I have fashioned the discourses around the beginning and the end of the most important thing in the world, which is the answer to the question: What is the most important thing in the world? You.

If you are interested in 2012 as a prophetic event or as a Mayan calendrical event or you viewed one of the cinematic 2012 guesses, you may already have an opinion about what 2012 means as a global or even galactic event. Perhaps you already know the Mayan Calendar is the end of a very long Bak'tun or "bundle" of time. If you are reading this text of discourses on 2012, you may have an interest in some area of Mayan Cosmovision that comes from an academic or personal interest. You must then first know that I am neither a Mayan nor a Mayanist. I am a United States Mid-westerner with a deep Welsh heritage. I am an American. In fact, Mayans are Americans too as most Mayans live in Guatemala and many live in Southern Mexico, the Yucatan, Honduras, and Belize. I have spent five years in Guatemala rather on and off which is certainly not as long as many anthropologists, ethnographers, linguists, or culturists; but longer than many of them. I will say that I did not simply wile away my time as a tourist. I studied Mayan culture, interviewed Mayan people, and attended many Mayan ceremonies and other Guatemalan religious ceremonies, both Catholic and Evangelical. My interest was not that of the usual academic hunters and gathers of Mayanness. Nor did or do I have any interest in trying to prove the Mayan prophecy or Mayan way of life as either true or superior or inferior to modernization. People far differently qualified than I have already attempted these arguments. In fact, I will prove that the Mayan calendrical system is not a prophecy at all, and that Mayans, as a people, are no wiser or different from all ordinary Americans. It is, in fact, these two truths make the Mayan Cosmovision imperative and why this vision is the real answer to why you are the most important thing in the world.

Why would I seek the answer to this question from Mayan cosmology? Why not simply use the popular ideology of religion or psychology or humanitarian motivation or even science? I hope, in the end, you will not say that I have proposed a bunch of wacky ideas, but I am going to offer you what may be one here. The land speaks, and we will now be directed to listen. We have never listened. Why? In great part because we are rather interlopers on the lands of the Americas. As Americans we are unique among the peoples of the world. Anyone can become an American. This is not true of any other nation in the world. A cadre of people from just about everywhere populates the North of America, and the Central and South of America is an amalgamation of true hybridity. Any casual traveler can point to differences in the Americans North, Central, and South, but l would like to suggest language is the only notable one in the end. What I would like to note is an imperative obvious sameness, namely that all Americans inhabit the American Continent. Ok, now the wacky part. The voice of the unbroken American land must become the voice of the people. Since Americans are mostly not indigenous to the American Continent, they have serious trouble hearing the voice of their exceptional land. Americans have either superimposed a foreign voice on the land or they have suppressed the land's voice through their colonial military and religious domination. Before you think that I am going to criticize religion or politics of the Western World, I want you to be assured I am not. Indeed, in the end I will support them, all of them. We must support them because it is in their plurality that makes unique their unity. But for a bit here let's give our indigenous people a chance to speak where others have failed by returning to "you" as the most important thing in the world.

I already knew the answer to this question of your importance, but I could not explain why my answer was the correct one until I coalesced Mayan Cosmovision with other religions, no matter how popular or parochial. When I posed the question of what is the most important thing in the world to the many groups and individuals I have, I always received an array of answers: love, Jesus, Mohammad, Buddha, God, various religions, children, spouses, parents, freedom, and pets are the most frequent responses, and that is the usual unscientific order. In fact I would put my dog first. These are very nice answers, and I believe they all point to the real answer. Typically, when I tell my audience the correct answer they say that I have tricked them, or that I am wrong. Oh, yes, there are those who say my question is unfair because I ask "what" rather than "who." How can I prove that "You" are the most important thing in the world? Actually you prove it for me. Unless there is something tragically wrong with your thinking about yourself, you know how much you value "Being" and being you. You have done everything you possibly can to live as well and long as you can. Yes, you mess up. You may eat badly, smoke, get angry, are unkind, or take other odd risks. This only proves you are human not that you don't think yourself important. Also, I am not talking about your self-importance as a matter of ego or hedonism, obviously. Further, I am not talking about your inability to put yourself before loved ones or country. I can assure you I believe I would sacrifice myself for my children. But this only proves how much we care correctly about ourselves, as our loved ones and our country are really us. Think about it honestly. Won't you do everything possible to care for the most important thing in the world, you? All the sane peoples of the world believe this about themselves.

Still, what in the world have the Mayans to do with the answer? Aren't they simply a very small group of indigenous people who used to like to sacrifice other Mayans to some god until Spanish Catholics saved them and now they live mostly in poverty and can't even drink their own water? If we want to be cultural critics, we can think of even more criticisms. I have lived with and among Mayan people and I could easily be critical of them in many ways. But, I can be equally critical of my very fine neighborhood in Columbus, Ohio, or many places I have visited and lived in my own country. Being negative is a very easy thing to do. I am sure that this is why it is so popular. I am never interested in problems be they personal or world. We are all problems and we all have them. What interest me are solutions. I want to live by the idea that there are no problems, only solutions. The Maya Cosmovision is in great part a solution.

Mayans are the largest group of indigenous peoples in the Americas. That is, they are people who belong to this land as original not interloper. What does it take to be Mayan? In my conversations with them, there are three distinct qualifications that make one a Mayan person: 1.

Born as an indigenous Mayan 2. Know a Mayan Language and 3. Believe in the Mayan Cosmovision. It is the Mayan Cosmovision that we all can share. Does this sharing mean we become Mayan believers or adopt some kind of Mayan religion? Not at all. The Mayan Cosmovision is not a religion. In fact, most Mayans are Catholics. Therefore, saying that we believe in or embrace the Mayan vision does not make us Mayan, and it certainly does not make us Catholic or even religious. The Mayan Cosmovision is captured in the Mayan calendrical system. In case you are not someone who has had some deep or even passing interest in the Mayan culture or the 2012 hype, I want to summarize the Mayan Calendar and its 2012 Cosmovision. The Mayan Calendar does what all calendars do, mark time. Although there are quite a few guesses as to when the Mayan Calendar begins, no one knows for certain. Still, most agree that this Calendar ends where it began-- December 21, 2012. Yes, this is the 2012 winter solstice.

You can easily find extensive information on the Mayan Calendar and its complexity by reading any of the many books, pamphlets, or web sites about the calendar. Here, I will give you a brief overview, if for no other reason than to satisfy how the calendar works in a very simple fashion. Actually there are three calendars in Mayan use— the Haab, the Tzolk'in, and the Long Count which is a combination of the Haab and Tzolk'in. This calendrical system is a sophisticated refinement of even earlier Mesoamerican calendars. The Haab calendar counting will be most familiar to those of us who use the Gregorian calendar. The Haab consists of 360 days of time activity plus 5 rather do nothing days to make a logical 365 day solar year. The Tzolk'in is a 260 calendrical counting of spiritual days. Time in the Tzolk'in is marked by days not months. It is the calendar used by Mayan Shamans or more properly called Day Keepers to ascertain ones birth day glyph and its symbolic and mythological foretelling. The final calendrical counting, known as the Long Count, is a world time counting and theoretically is rather a backward count from December 21, 2012 to discover the beginning and the end of time bundles known as Bak'tuns. Since the Mayans used zero, they were the first in human history to do so, perhaps the count is off by a year. In the end, if you are interested in the "business" of time, use the Haab. If you are interested in the "astrology" of time, use the Tzolk'in. If you are interested in the "end" of time, use the Long or World Count. Going against the grain of far too many "prophetic" meanderings about the Long Count's end of time, I will tell you that I agree with the scholars and Shamans who know that the 2012 date is in fact the end of the thirteenth Long Count time, not the end of time, the world, or some apocalyptic event, but the resetting of the Mayan Calendar.

Yet, this calendar has been hugely complexified, or resignified, by authors who have written very long books about it and especially about its seeming vision of the end of time, December 21, 2012. I have read all of these books. They are all interesting, and I would be the first to congratulate most of these authors on their difficult work, but they are all wrong. They are wrong in two ways: 1. They believe the Mayan Calendar is prophetic and 2. They

7

believe the Mayan Calendar marks the end of the world or at least the end as we know it. I know some Mayans who believe this too, but only the ones who have somehow directed their guidance from "astropsychologists" or New Age foreigners who are clearly not Mayan.

The idea that the Mayan Calendar's Cosmovision is prophetic is predictable. If you take the time to make even a terse study of the Mayan Calendar and the many writings about it, you will see why 2012 is an important end to a calendrical period for the Mayans. What will happen around December 21, 2012 (or some near date depending on the foreign forecasters view) is the end of the Mayan Calendar's final or 13th Bak'tun, that is, the end of time. What will happen when this "time" ends? I can assure you none of the "prophetic" meanderings that have become so popular in apocalyptic literature and movies about 2012. What do the Mayan Shamans (literally, those who know) believe will happen at the 13th Bak'tun's end? Simply the calendar will be reset to zero, a number the Mayans had before anyone else. A brief study of the Maya Cosmovision will reveal that most Shamans report and exhort the earth's awakening in 2012. Nevertheless, this is a bit of an error. 2012 is about our waking up to the land not to the land waking us up. 2012 is a stage date. It is a year bundle of time that will give us the opportunity to exercise our free will and chose a path we have never before chosen. 2012 is not a prophecy but a prefiguring, anticipation or heralding. It is a time stage to glorify the most important thing in the world: You.

What about you and 2012? Maybe you won't even be here in 2012. You may be transformed for the final time before 2012. In fact, what 2012 is about is precisely that: transformation. You do not have to wait for 2012 for this transformation. 2012 will happen with or without you as will 2013, 2014, 2015, and tomorrow. Do you, that perfect and imperfect you, really want to transform? Transform into what? Transform how? Let's not wait for those answers. The transformation is raising you to your rightful position: you as the most important thing in the world. You not believing but knowing that you make all the difference, and difference making is up to you. Let's consider what really transforms us in the rest of our discourses. But let's first consider what the prophets of the prophecy prophesize and why they are wrong.

I am not in the condemnation business here. The books, pamphlets, web sources, and seminars about 2012 are mostly serious, intellectual, or hopeful testimonies about the transforming power of 2012. In fact, the vast majority of criticism leveled at the 2012 authors of prophecy is not nearly so compelling as the "New Age" hopefuls. The critics are mostly Christian religious folk who essentially argue that their apocalypse is better than anyone else's apocalypse. Sometimes a "scientist" or two will weigh in with such dynamic comments like, "Well if I want to be a better person then I can choose to be. I don't need some world changing vibration to make me." These folks are rather like the novel *Lord of the Flies* science character Piggy who believes that science could actually rescue the little savages from their island prison, never considering the imprisoning darkness that lay within them. Nor are the 2012 prophecy authors of 2012 knowingness confined to minority groups of "New Age" spiritualists. Let's begin there. In his wonderful and clearly the best work on Mayan epigraphy, *Breaking the Maya Code*, the renowned archeologist Michael Coe recounts this experience while working on Mayan Glyph reading in Palenque, Mexico:

> To the southeast of the Palace (Palenque) is the Cross Group, dominated by the Temple of the Cross, so named because of the cruciform world-tree found at the center of the relief in its temple-top sanctuary. The Maya, both ancient and modern, have had many curses laid upon them, and the fantastic theorizing by the lunatic and near lunatic fringe is one of them. The Temple of the Cross relief has been a frequent target of crackpot notions; back in 1956 my wife and I sat in a Mefida café next to an American who first identified himself as An Apostle of Jesus Christ of Latter-Day Saints (reorganized) [Mormons], and then assured us that Jesus had returned to earth after the Crucifixion and preached to multitudes from the Temple of the Cross (Coe, 193).

Therefore, Christ was there and anointed the Western World, that is, Mormons, with the true way of this new "chosen" people. Of course the world cross is nothing of the sort. Actually this is a temple monument to the rulers Pacal and son's coronation. The World Tree is a central symbol in Mayan mythology and epigraphs. The Mayan world tree is a representation of a popular indigenous tree, the ceiba, in Guatemala. It is an important symbol of life that flows through the earth both unseen, roots, and seen, trunk and appendages. It has nothing to do with Christ's crucifixion.

Speaking of the world tree's appropriation, let us turn to the best book on Mayan Calendar prophecy, *The Mayan Calendar and the Transformation of Consciousness,* by Swedish author Carl Johan Calleman, "Human beings do not create morphogenetic fields—the world tree and world mountain do (Calleman, 192)." Strictly speaking, morphogenetic fields are cell groups that dictate the formation of an organ group—for example, the cardiac cell group will generate a heart. The idea is that cellular biology results in something rather useful. I suppose Dr. Calleman's notion here is that somehow we can create cellular fields through some thought, or more likely a spiritual process, beyond the biological.

Indeed, Biological transforming is in fact our most important transformation. It is real. It is so real that it is the habit of "transformationalists" of all stripes to attach the spiritual to the physical, i.e. are you "born again?" We will take up the importance of biological transformation in Discourse Two. Calleman extends his morphogenetic meaning by hitting on a nice note about the fact we will have some choice about joining the 2012 galactic transformation, "We do seem to have free will in deciding whether we want to surf on the waves of creation. There is little to indicate that our choice is predetermined or that there are any 'chosen ones.' Nor would it matter what religion we believe in (Calleman, 192)."

11

This is why I rather like Dr. Calleman's positions. His desire for a physical solution to our seeming spiritual problems is off the Mayan mark—"surf the creation waves" indeed--but he is consistent in his thrashing religious egos. Still he is too interested in the Mayan Calendar's ability to be what it is not, as he explains, "The Mayan Calendar, an analytical tool generated by the Western Hemisphere, remains our most important instrument for studying the cosmic plan (Calleman, 197/9)." What exactly is the "cosmic plan?" According to Dr. Calleman, this is a plan of enlightenment, "The transition to the Maya Calendar's rationale is that it points out the path toward enlightenment" (Calleman, 214). Evoking the titillating vocabulary of "cosmic plan" and "enlightenment" are like those infomercials that spend an hour telling you nothing but tempting you to think maybe, just maybe, this will work for me. You are already in the cosmic plan; you are here. You are already enlightened; you know that you are the very image of god. Now you must imagine how important that is.

One last word from Dr. Calleman, in speaking of the Mayan underworld, Dr. Calleman misses another important physical reality by hoping for a spiritual overtaking of reality, "The Galactic underworld is more about telepathy and intuition than technology . . ." (Calleman, 140). Telepathy is with us already and it is with us through technology; you may twitter that if you like. By the way, I recommend Dr. Calleman's three wonderful books on 2012 above all others if you insist on needing a fairy tale 2012 that sounds like science. I do this because you might otherwise find yourself reading such crap as the following:

> As an exo-biologist, I have been schooled in the many aspects involved in transforming your present body into a fully-conscious one, or a 'New You'. To birth this 'New You' involves a process that presently is integrating your Spirit (Full Self) with your physical self. A unique change in your physical body's RNA/DNA and in its current energy centers (or chakras) is occurring. These remarkable modifications will permit you to transform yourselves from your current state of limited consciousness to a state of full consciousness (Vasiliy Kuznetsov http://2012-year.com/)

"Full consciousness" will be further explained by Kuznetsov as the merger of spiritual and physical consciousness. This is the most tempting and timeworn misuse of our spirit--this earthy search to make our physical ethereal. This is the error that has made all religions relevant and dichotomous. When these physiospiritual attempts are made from the likes of a self-proclaimed exo-biologist (the branch of biology that deals with the search for extraterrestrial life and the effects of extraterrestrial surroundings on living organisms), we can only be amused at best. Let's look at one more:

> Our planet steadily comes nearer to transition in 4 measurement, and a problem of all people - to have time to prepare itself as spiritually, and in respect of transformation of the biological body

for successful assimilation in new measurement . . . At first you will feel, that there is something not clear. At you [sic] not clear indispositions, a fever will begin; process of opening of supersensual perception, that is, mass opening of the "third" eye will begin. At the first it will be felt by simply very pure and light people. Then you will start to see structure of space which will change at [sic] you on eyes . . . http://2012-year.com/home/scientific-and-public-theories-about-2012.html

I don't even know what to say about such absolute silliness. What is in any case relevant about any of these attempts at 2012 prophesying is three fold: 1. Using 2012 to promote some transitional point in time 2. Posturing that there will in fact be a world or even galactic transformation and 3. That the transformations will be some sort of physical or biological transforming that will result in a spiritual transmutation, that alas something will happen to change our biology and forge us into a spiritual being. This is the denial of our physical as our spiritual. We just can't accept our spirituality because our belief systems have so separated us from our spiritual image. We forget very soon in life how important we are, that we are the most important thing in the world. We learn to fight our spirituality by becoming so many things other than who we were created to be. Lest that sound overtly religious, let's turn in Discourse Two and the importance of biology to transform us into the wonder of our importance.

# Discourse Two:

# Biology, The Time of Our Lives

Our biological transformation is powerful. This transformation is so vigorous that some of the apocalyptic authors of 2012 are even tempted to believing that it is indeed not so much transformation but a hopeful transmutation, a way to become something or someone else, another "creature" if you will. It is not. Our physical stages of life—from baby to child to adolescent to adult phases—are so recognizable and so real as to become a metaphor for all our other transformational beliefs. In fact, it is in our biology that exists our most powerful spirituality. Our physical condition is our spiritual condition. This biology, this consciousness, is the most important knowledge system that we possess. It is so obvious and so important to us that we call it life. It is without any reasonable, psychological, illusionary, or spiritual argument the reason we know that we are the most important thing in the world. Still, for all of us it has two essential limitations: 1. Our physical or birth limits and 2. A seeming limit or partition from the spiritual community.

Our physical limits are embodied in our physicality. This knowledge tempts us believe our physicality also limits our spirituality. We know we are limited and at the same time are exhorted to believe we have no limitations. Understanding our limitations and our potentials is vital to a life of joy, success, satisfaction, and self-importance. I have been a teacher for most of my adult life. I have done many other things, but mostly I have been employed as a teacher. I know I have been a good teacher by my satisfaction with doing it well and by what others have said about or rewarded me for in my teaching. However, I was not "born" to be a teacher. I am sure there are many other things I could have done with the kind of physical limitations and potentials with which I was born. I am sure I could have been quite a good attorney, judge, politician, minister, or factory worker. Still there is one employment in my life that points certainly to my limitations. I am a soccer coach. I have coached high school and college soccer for over twenty years. I have experienced only one losing season in all those years. I am good at it. But what I am not good at is playing soccer. I did not grow up with it. I played football, and I was not good at that either. But what if I thought I was good at football? What if I had dedicated my life to being the best linebacker in professional football? I can tell you that I would have utterly failed. Why? I do not have the physicality for it. I could never have gained the size, strength, speed, or quickness that is demanded of any good athlete. Yet, I see people all the time trying to be something their birthright will never allow them. They try to go to a college that would make them miserable given the other students' board scores at that college. They try to work at a job where success eludes them constantly. They remain in relationships that make them unhappy. They refuse to accept their wonderful birth talents or even lament them, and they may work their whole lives to undo what is so them. I want to submit that all of us recognize how heroic it is when someone overcomes what seems to be a physical limitation and succeeds at some wonderful or personal vision. I want, though, to submit that the truly heroic lies in the pursuit of our given physicality. Our physical is our spiritual. You are your spiritual. Whatever do you think is meant by our creation in the image of God?

This image, this imagination, is the reality that we too often do not want to accept. The factual transformational reality of our biology is the temptation of our belief that maybe, just maybe, we can be transmuted into something different from or physicality, indeed, something spiritual. Since this transmutation cannot factually occur, we create myths. Many of the books, seminars, and writings about 2012 spend a great deal of time creating Mayan myths. Let me say an important thing about myths straightaway. Myths are true. But you already know this. I have been teaching myths for many years. When I ask students to define myth, an inevitable and usually a first answer is that a myth is something that is not true. I then ask students to tell me about their life from birth to the present. Suddenly the myths start rolling. We have one area of our physicality that is completely mythological: our infancy, and we are going to have another one, our death. Neither the baby nor the corpse have the apparatus to tell us what it is like to be a born or to be dead. But lots of people tell us what it is like. These seers know what is before life or after life. They report tales from the yet to be born and the dead. These people are all fakes. If you want to believe in these magical realities, fine. I must admit that I often want to. But believing in something and knowing something are very different. This distinction is so important that Christians, in fact, are not asked to believe in Christ but to know Christ. Here too, the Mayan Cosmovision weighs in on the knowing rather than the believing. The Mayan Cosmovision is a very intellectual use of a reality, that is, of the physicality of time: a calendar.

Time is very real. I have friends who always say that life is an illusion and that there is no reality. Some of these friends are physicists and academic phiolosphers and others are spiritualists. We should then ponder that if there is no reality, then what is unreality? As intellectuality sporting as the philosophy is that time and reality do not exist, it is also as, well, wacky. We know what is real. We are. Remember that you are the most important thing in the world, and you are your reality. It is in fact our physical reality that creates who we are at any progression of our biology. We mark these progressions by time. Time supports our physicality. That is why we use it. Divine time is not available to us. The Mayan Calendar is unique in its time marking. It is a calendar that marks time in physical terms that represent physical solar movement, 365 days. Not only was this useful to the Mayans' capital agricultural life, but as well to our sense of capital life. The Mayan Calendar also marks a kind of spiritual time, 260 day signs that offer our positive and negative birth day potentialities. At last the Mayan Calendar marks time in large bundles, Bak'tuns, some 400 year periods of beginnings and endings. December 21st, 2012, is the end of the 13th Bak'tun. Since 13 is in fact a good luck number for the Mayan— for instance turtles are rather sacred creatures because of the 13 scutes on the turtle's carapace or outer shell. Because of the sacred use of number 13, it seems logical to some that the thirteenth Bak'tun's end will be the end of time. Of course it is just as logical to assume that the end of the thirteenth Bak'tun will reset the calendar and begin yet another Bak'tun or start time again just all calendars do. Nevertheless, the end of Bak'tun 13 signals the opportunity, sets the stage for a new beginning. It is not so far fetched to believe that something other than you will cause a special change or a transformation. You know, let the gods do it, or science, or galactic vibrations, or maybe aliens.

Inevitably so many authors of Mayan archeopsychology end up offering such nonsense. They also soon attach very physical terms to the discussion to try for some form of validity. They jump on the psychology of "stress" and "burnout" relief as well as the end of dualism, good and evil, by way of a world or galactic transformation of frequencies that rids us of all kinds of maladies. They want a miracle or magical transformation to make us spiritual beings rather than physical beings. They don't believe that you are capable of transforming yourself. They don't believe in your importance. They, like too many of us, believe we need to be rescued from our birth condition. They believe we are not spiritual in the physical and that we cannot be spiritual beings as long as we practice physicality.

Let's look at this more closely. Are we physical beings? Are we spiritual beings? Are we both? I have made the case that we are only physical beings, and how important this is to our spirituality. A certain tenant of the 2012 arecheopsychologist and MayanOther spiritualists is that the Mayan Cosmovision either abrogates or will end duality.

I find this odd at the very first level of apologetics. How can they be in the 1. physical and believe in the 2. spiritual and not embrace duality? I understand the desire to make the case for moving the physical along to a spiritual revelation. I agree. What is this spiritual revealing that we all long for? How can we move to a spiritual level of consciousness while trapped in the physical? We do this by our making duality unity through consciousness not unconsciousness. As long as we look to a formula that demands outside or fabricated events to transform us, we will continue to fail at transformation. Einstein once mused that we cannot solve problems with the same consciousness that created them.

Let's then begin with consciousness as the most real thing we know. Reality knowing is not only the most reliable knowing; it is the only knowing. Conscious spiritual knowing, reality, is therefore separate from unconsciousness spiritual knowing, unreality. Spiritual knowing does not exist and can and will not exist apart from consciousness. Consciousness is reality. I think of all the philosophers and psychologists, Shakespeare knew this best. Poets usually do. It is in fact the job of poets to make our myths reality. Hamlet's most popular "To be or not to be" soliloquy is one of the best moments of Shakespeare's realty making of myth. "Thus conscience doth make cowards of us all," Hamlet's ending line of the soliloquy is one of the most relevant statements about our physical lives. If we look beyond, as Shakespeare certainly did, the speech as suicidal, we see the poet's meaning of life. We do not know anything but what we know. Life before life and life after the life we know is in fact "an undiscovered country from whose borne no traveler returns." No baby and no corpse can tell us what his life is like or what it was like to be before or after his conscious life. As E.M. Forster, another fine poet, tells us in his wonderful work *Aspects of the Novel*, only fiction can create these lives after and before conscious life.

What proof do we have that consciousness is so important? That is, more important than all the spirit constructs of all religions, cultures, philosophies, or prophecies. We all know the cliché of taxes and death. I will leave taxes discussions to television. Death is the reality that holds true for all of us; so much so that even discussing its inevitability is a waste of time. When we see death we mostly react in one of three ways. We are not much emotionally interested, as we do not know the dead. Go through the obituary pages and see if you emote over the people there. I personally never look at obituary pages, but I had a friend who said he always looked to see if he had to go to work or not. We can be horrified by some deaths of those we do not know when we see the unjust or horrific violence that caused a death or deaths. Last we know, that is, we are conscious of a death. We knew the person, and we are saddened, deeply or remotely. Why? We all know death is coming. We mostly believe in some form of spiritual moving on or place for our dead and death. Is it therefore unscientific or unfaithful of us to mourn? What in fact are we mourning? Part of our lamentation may come from the manner of death: a child's death, an accident, a despot's work, a horrific event. Yet, our deepest mourning comes from the end of a consciousness. This is Shakespeare's meaning. We know that the physical, conscious being of the corpse is no longer with us. In a morbid way we could keep corpses with us, preserved in some chemical stew. But this is not what we mourn. It is not the loss of the fleshy presence we miss. It is the reality of the person we want back. It is the reality of us, our consciousness we cherish. "Thus consciousness doth make cowards of us all."

Does this cause a dismissal of spirit and religion as fictions? By religion here I mean any organization you can think of that seeks to tell us that there is a before and more likely an after life. Or to put this another way, any organization that believes in duality: heaven and hell, good and evil, life after life, life before life, reincarnation . . . By spirit I mean a soul that is separate from our physical or dwelling separately in our physicality. Indeed, the Latin root for spirit, *spiritus*, meaning breath, certainly does exist as a kind of unseen spirit. Still spirit as breath is quite physical and very provable. I attach these two, religion and spirit, because they are handmaidens in most minds. Of course there are those who are religious without a hint of the spiritual and those who are spiritual without the slightest attending to a religion. But religion isn't really the point, is it? Nor is science or medicine or philosophy or psychology or any of the belief systems which become dependences in trying to shuffle off who we are and becoming something else. These manifold methods of searching out some transformation that will end who we are in time or who we will be in timelessness. Struggling somehow not to transform ourselves but to be transmuted into a different and what we hope will be a better, or more complete, or alas, spiritual being. Human history is filled with these many attempts and inevitable failures at destroying who we are. What we are suggesting here is that these have failed and too often result in destructive thinking of others as well as ourselves.

Let's make an important amplification of what happens when you make yourself the most important thing in the world. You as the most important thing in the world must come to realize that all you look at is the image of you. When you have transformed the love of yourself into a world mirror, you then actualize yourself as the most important being in the world. Sure, love your neighbor as yourself. Just how hard is it to accomplish this overworked proverb? Certainly we must admit it is hard for us and seems damned impossible for the world altogether. Is 2012 just another time to have a go at it again? Can we, can the world, really change?

Here it is necessary to ask if transformation is the same as behavior modification? That is, will you behave differently when you are transformed? Again, the answer lies in our biology. Just as my biology does not permit me to be a great athlete, it also inhibits my ability to be a different actor in many ways. I can become so many unloving things. Rather than indict myself, I'll let the Mayan Calendar do it. My birth day sign is Ajpu'. Accordingly, I have some very nice positive traits and some very awful ones. On the good side I am judicious, valiant, brave, good, kind, a victorious warrior, a hiker, a scientist, and/or a farmer. On the vile side, I am judgmental, quick to anger and unforgiving, vengeful, and aggressive. Furthermore, note the duality inherent in my "oh 2012 will end duality" personality. Also note, according to Shaman Daykeepers, I can choose among these traits. Ok, how true is this? Well, I don't really know. I would say that I am judicious and judgmental as well as kind and unforgiving. I am certainly not a hiker. My idea of camping out is checking into a Hampton Inn. I have little to no touch for the math of science. But, I have always been attracted to farming. In fact, one of my few memories of being a child was my first grade teacher's assignment to draw a picture of what we would like to be when we grew up. I drew Jesus on a tractor. I just couldn't decide in this early career pathways teaching if I wanted to work miracles or plow fields. Our society makes it quite a lot easier to follow an agricultural career than a savior one, so farming it was. But while hoeing tobacco fields at age 17, I rubbed the sweat off my brow and decided the savior gig might be easier. Teaching is sometimes close. Point is, my day sign is at least accurate on the farming guess and that I could make choices.

So is the answer to transition simply to make a choice? No, our biology will not permit it. I want to repeat how important it is to accept our biology and to embrace ourselves—all of ourselves as the image of god—as the most important thing in the world. Our biology is our spirituality. Making behavioral choices is difficult but possible. We can and do quit smoking, lose weight, change relationships, modify our actions or not. Transitional choosing is not possible. We can't stay babies or children or teenagers or remain 39 or stay our death or become a spiritual being. Biology and time will not permit it and it is time we admit it. Our biology is our spirituality.

If this is true, why talk about transformation at all. Should we be as Hamlet suggests, "What is a man, if his chief good and market of his time be but to sleep and feed? a beast, no more." What do we mean by changing the time of our lives? This in fact is the current interest in the 2012 change phenomena. Again, we know our biology is immutable. But we can't let this turn into an illusion either. Typically this is manifest as the illusion that all our ills are biomedical or reasoning that the ills and evils of humanity are strictly evolutionary. This can become an easy excuse for surrendering to a rather helpless inaction to repair suffering, cruelty, poverty, genocide, and ignorance. We want, we believe, someone or something else has to do it. God and government and evolution have not done it yet so at an appointed time the whole of nature, the Galactic community will rise up and make us all good folk drunk on new interstellar vibrations. Any one who believes this, and there are plenty, will be terribly disappointed on January 1st, 2013.

Are we in fact "a beast, no more?" Are we helpless to replace duality with unity? No. The transformation is possible. How? Remember you are the most important thing in the world. Knowing, not believing, this is the beginning of the transformation. And this is a world transformation, not just a you transformation. If you want to change a behavior, you are welcome to have a go at it. If you succeed, congratulations and celebrations. If you fail, just remember you were limited by your biology and you should embrace its wonderful success at keeping you who you were meant to be, that is, the most important thing in the world. Notice I have never said, "The most important person in the world." Let me tell you why.

Remember 2012 is a prefiguring event, not a prophetic event. Let's think about that. Prefigure is an action verb that means to suggest, in form or likeness, a person or thing or event that will come later. More directly to prefigure is to imagine a person, thing, or event in advance. What is important about this imagination is that it is not a wild guess. It is to imagine the future with evidence, typically evidence from the past. When the Mayans conduct a ceremony, always with the calendar in mind, they count on the ancestors to guide their conduct and thinking. This concept is best related by Miguel Leon-Portilla in his *Time and Reality in the Thought of the Maya,* " . . . All thing(s), good and bad, have their appointed time to end and to be repeated in an endless cycle, in which past is prophecy and prophecy is the past (Portella, 185)." For the Mayan, prophecy is not some fantastic future prediction but the very act of prefigurement. It is the ancestors who brought them to life, biology, and it is the ancestors who gave them time, the time of their life.

What we do with the time of our life, with our real consciousness, is what creates transformation. You, the most important thing in the world, are a co-creator of unity. You cannot do it alone. There are no supermen, deities, or galactic phenomena coming in 2012. They are already here. You are they. As the Mayans know, they have been here before, "our ancestors inform us." Johan Calleman says, (b)lind to our own consciousness, we fall prey to the illusion that it is the material world that is real (106)." He is right; we can be blind to our own consciousness. We can deny the wonder of our biology, our real consciousness, our reality. He is wrong to suggest this reality is not material. Our failings may be proof of our materialism. But it is in fact our failings that inform us that we can succeed. I can modify my behavior. It is a choice. I cannot transform the world, but we can.

Remember what we said about the definition of spirit, *spiritus* meaning breath. Our biological breathing, invisible in and invisible out, is a metaphor for how our biological is our spiritual. It is also a powerful way to act on our desire to modify behavior and transform our world. Think about breathing in as taking care of yourself, and breathing out as taking care of the world. There are many aphorisms, proverbs, maxims, and clichés to live by, but few of them are actions. Breathing is an action that has a chance to help us bring harmony to ourselves and our world. It is an involuntary action that can become a thinking event. The point is this is one way to think about transformation. Because thinking and knowing are handmaidens, it is the thinking that leads to knowing that guides us be members of the transformational community. Before we look at the hopeful changes 2012 can bring, let's look at how transformational thinking can actually change the world consciousness.

## Discourse Three:

## Membership in the 2012 Myth

A discussion of 2012 is discussion that is certainly terminal. On January 1, 2013 the discussion is over. The calendar will reset, the 13th Bak'tun, bundle of time, will end and the myth will succumb to a new time. Remember in Discourse Two I introduced myths as both false and true. I concluded that myths are true. Certainly we can and do use the word myth to mean something that is not true. I would say, for instance, that many of the infomercials about diet, hair growth, quick riches or cures are myths, and I would mean that they are mostly lying. But how do I know this for sure? Lots of people in those infomercials give testimony that they are slimmer in body, fuller in hair, richer in money, or buoyant in health from these various procedures or products. Am I calling them liars? Actually, I rather am. But I have never met any of these people and I don't know anyone who has. I don't think of anyone who is not behaving as a criminal a liar, but I do question the discerning ability of people who believe in or testify for myths of unreality. I do not only mean this for quirky infomercials, but also for any belief system that is only supported by testimony rather than reality and supported by people who will not spend the time and effort to discern if something is, in fact, true.

Am I then saying that only facts are true? No! I am saying that myths are true, and it is in being a discerning person that makes the most important thing in the world, you, an active member of the 2012 myth. Is there a way to discern what myths are true? In fact, how do we know what is true? Isn't truth in the eye of the beholder? Isn't one person's truth just as true as another person's truth? This sort of philosophizing falls into the same discussion we had in Discourse Two about there is no reality and everything is an illusion. Again, this makes for exciting rhetorical philosophy, but accomplishes little as active transformation. How can we give legs to transformation? Will 2012 come and go and find the world much the same as it is today, mired in the duality of finding ways to vilify our own image? How can we make the transition to know what to do and how do myths make this real?

Recall that I began these discourses with what I called one wacky idea: The earth speaks and it is time for us to listen. I based this on the reality that almost the whole population of the Americas are people from another land or people who were subjects to the conquerors from another land. I said this is the real reason the Mayan voice is worth hearing. It is the *vox populi* of the Americas. There is nothing wrong with the other voices. All the countries of the Americas have a national and cultural voice. And I am not suggesting that those voices are false or unsuccessful. I could suggest that and indeed, I could successfully argue it, but this is not the purpose of 2012. I am also, and this may surprise you, not suggesting that we listen to the land as a matter of environmental ecology or mother earth spirituality. What the Mayans have to say about the land is that the ancestors are there and that they are available to guide us. They are there to guide us not as voices from the dead or spirits of the air. Their guidance is one of time, the most important quality of Mayan thinking. This is well noted by ethnographers, archeologists, and my own discussions with Mayan shamans:

> With their astrological preoccupations coexisted the knowledge which, through the symbols, had led to the discovery of a universe conceived from the point of view of an ultimate and all embracing reality, time. The relationship with the ancient myths was preserved, and need for prediction became fused with rigorous measurements and computations. So was born a most unusual religio-mathematical vision of the universe (Leon-Portilla, 107).

Let's mold this idea into action: all we have is our time on earth and we must learn from the past how to have a future. Indeed, this falls into the usual cliché about learning from the past so as not to repeat the past. But for the Mayan, what we learn from the past is not how not to repeat the past but that learning from the past is more of a cause and effect in that attending to the past causes us to repeat the past. This is also a very direct look at our own pasts and our tendency to be guided by what we believe are the facts of past influencing our present. It is in discerning our past facts (data), our mytho-facts (perceptions), and our myths (metaphors) that manifest a belief system that results in performance. Past facts and perceptions are our enemy because they are past knowledge and in no way represent our present reality. Myths are our future as they guide us forward. These myths are real knowledge, not self-delusional past reflections or other-perceived criteria that chain us to a life of duality. As the ancients of our land will tell us, our myths are true or false and we are responsible to discern and act on that truth. We are responsible to know, to measure the metaphor to make it real. This is the religio-mathematical. A "religio-mathematical" idea is a hard concept. Let's make it even harder. A myth is that which is not true that creates a truth that is false. For the Mayan it may go something like this: the sun moves eternally in a measurable configuration (fact), we were conquered and enslaved by other nations who superimposed their myths, religions, on us (perception), the eternality of the sun and its life engendering on earth where our ancients lie and where we will retire makes us an unconquerable eternal community of one heart (myth). How does this translate into reality of action for the Mayan people? It does so through the performance of ceremonies which are thanksgivings for many things and by their conduct of one heart kindness to others. The myth becomes an action that is real. The knowledge is repeated through ceremonial reminders and this results in the action of living joyfully through an enslaved past and an impoverished present. This may seem an oversimplification because it is. Still this is the knowledge which guides Mayans' life habits. Forming life habits is what we are talking about here. That is, thinking about how to live and acting on it.

This is not, by the way, easy. This is hard work. It is knowledge work. It is the work of reality that has yet to be real. How is reality not yet real? How, in fact, can we create reality? The answer lies in the art of our action. That is, the art that calls to action. Let's look at three examples of this, and so that you do not think I mean only pictorial or literary art, although I do mean those too, I am going to start with the art of Edward R. Murrow, a 1950's news anchor:

> We will not walk in fear, one of another. We will not be driven by fear into an age of unreason if we dig deep in our history and remember we are not descended from fearful men ... who feared ... to defend causes which were unpopular . . . The actions of the junior senator from Wisconsin have caused alarm and dismay . . . and whose fault is that? Not really his; he didn't create this situation of fear; he merely exploited it, and rather successfully. Cassius was right, "The fault, dear Brutus, is not in our stars, but in ourselves." Edward R. Murrow newscast on Joseph McCarthy

McCarthyism has become a word for the kind of witch-hunt mania available to our history. Murrow's newscast was not solely responsible for McCarthy's demise. A public willing at last to think critically, with discernment, overcame those who will always try to trick us into following the wrong voice. In fact, if you Goggle McCarthy today, you will find a raft of people still defending McCarthy. Why? As Murrow tells us, the "fault lies within us." How does this work out in our myth formula: Communism exists as a threat to the American way of political and religious life (fact), there are communists among us, indeed, in our government, in Hollywood, and around every corner (perception), Americans are free to believe in any philosophy they choose and no communists sympathizers are involved in taking over the government (myth).

What is this fault to which Murrow references by way of Shakespeare? Let's go back in our own history a bit further than 1952. In 1798 Brockden Brown sent his novel, *Wieland or the Transformation, an American Tale* to then President Thomas Jefferson. We don't know if Jefferson read the novel but he did pen a letter to Brown thanking him for the book. This is a short novel without much critical acclaim. I believe Brown wrote this novel as a warning about the potential misuse of the awesome freedoms for Americans that Jefferson is so famous for proposing.

The novel goes like this: a brother and sister see their rabidly religious father die by spontaneous combustion. The brother and sister, Theodore and Clara, become fascinated by a combination of religious and rational philosophies. Both marry friends who are equally engaged in philosophical meandering. All the characters are seekers of truth in matters of life and after life. If only God could or would tell them, does God exist and is there an after life and how do we know? And now comes Carwin, a very strange character who is never very clearly defined—by the way, that lack of clarity and things like spontaneous combustion make the novel a boorish read. Carwin is a biloquist, a ventriloquist who can throw his voice and sound like other voices. This is a fictional idea invented by Brown. Carwin mostly uses this voice to escape trouble or engage in affairs. The novel's turning point is brother

Theodore's belief that the voice of god tells him to kill his wife and children, which he does and so much for religion and rationality. Did Carwin tell Theodore to kill his family? Did God? Or, more likely, does Theodore imagine the voice that he so much longs for to know the truth, a truth that leads him to horrific consequences.

Ok, what's the point? Why did Brockden Brown send the manuscript to Jefferson? I believe this little novel to be one of the most important works in American history, and I know why he sent it to Jefferson, the father of the idea of all people having a voice. Brown worried, as Hamilton and Adams warned, that people would not be able to discern among the voices. That both religious fever, born in the Puritans' breast, and rational humanism, born in the Founders' beliefs, would both resort to voices the public could not, would not discern—a country forever reviving evangelists and politicians that seemed much the same. Brown feared the simple person, Theodore, would not be revered for his wonderful contributions and equally the intellectual person, Clara, would never be understood. Alas, we would become a country following a dream of knowing without ever coming to the truth of unity. A new nation made up of a people always ready to be deceived by voices of religious zeal or intellectual rationalization—a country always in the voice of regress because of its very progress--a nation of duality rather than unity. Unity can only be manifest in reality, as following illusions is the very essence of duality. It is illusionary thinking that creates heavens and hells, elites and bumpkins.

The Mayans' hope for unity is the same as any nation's hope for unity, but it is based in a different paradigm. Again, as the author of the most turned to book on Mayan reality, Miguel Leon-Portilla, points out, "The Mayan view of the myths of time and space was represented as reality not (illusionary) mythology (85)." They saw reality as mathematical fluidity that represented reality over illusion. Leon-Portilla further explains how this reality worked in reality, "Essentially that cardinal points are not as fixed quadrilateral representations as Europeans thought but more fluid as represented by sun movement as fluid. Sort of an it's there going. Things are in motion more than fixed in place." Leon-Portilla reports that this was as practical a view for politics and religion as it was for time and space (196-200). This is the Mayan idea of the landed ancestors speaking. Create gods, create spirits, create heavens and hells, loves and hates, religions, . . . but do not live in their dualities without discerning the dichotomy that is inevitable in every one of them. This is the constant message of art. You, the most important thing in the world, cannot create an image of you without an image of others. The ultimate duality is not so much believing other people are not like you. It is believing that other people are wrong and you are right. I am not speaking here about the wonderful spirit of argumentation and persuasion that helps us support or change our ways of thinking about thinking. Let's look at another artful lesson to explain this wrong versus right kind of thinking.

Herman Melville's *Moby Dick* is sometimes grudgingly called a candidate for the great American novel. I have taught the novel many times. Students hate it. The language is difficult in its magniloquence and the novel is quite long. It is also about hunting whales—hardly popular these days. But mostly *Moby Dick* is difficult in its thesis: To live you only have to die. This is really a pretty popular theme in much of literary art and certainly it is a central theme in all religions, but I don't want to talk about this central theme. I want to point to one chapter, "The Monkey Rope." *Moby Dick's* main character, Ishmael, teams with a very unusual fellow whaler, Queequeg. Now Queequeg is a seven-foot, tattooed from head to toe, powerfully muscled African who worships a small wooden idol tucked in his knapsack. The Christian and soul-searching Ishmael rooms and sleeps with this pagan and suspected cannibal giant of a man. They become fast friends. As they are paired together in cabin so are they in every whale hunt. In the "Monkey Rope" Queequeg is stationed on the back of dead whales tenuously strapped to the *Pequod* in order to parse them for oil. This is a deadly activity as butchering a dead whale is slippery business. It is also a delightful business to the many sharks swirling to the kill. If Queequeg slips into the water, this would also be to the sharks' delight. In order to keep him safe, Ishmael is on deck with a rope tightly about him on one end and about Queequeg on the other. It is Ishmael's job to keep Queequeg steady on as he parses the whale. This is indeed a gruesome scene. As Queequeg goes, so goes Ishmael; as Ishmael goes, so goes Queequeg. The rope, the monkey rope, is the lifeline and potential death line for them both. This is an interesting bit of knowledge about whaling and a bit of an interesting story, but that is not the point. "The Monkey Rope," like all of *Moby Dick* is a metaphor and I think a rather obvious one. We are all on one end of the monkey rope. Every movement we make is one that may endanger or save another. This is an awesome responsibility.

I have given you three art examples out of thousands that inform us of how to act as the most important thing in the world. Like I mentioned in the beginning of this discourse, knowledge work is hard work. When we hear the voice of our leaders, of our religions, of our philosophers, of our scientists, and of our art; how do we know what is true? Here I want to borrow an idea about knowing from cognitive psychology. Since I have suggested that knowing and therefore acting on what we know is hard work, we can turn to the reasons knowing is hard and how our performance comes from our knowing. In Discourse Two I discussed how biology makes much of who we are. Here I want to turn to the psychology of our actions. Apart from our biology, we learn from our environment, what we see. We put that seeing in our working memory and store it in our long-term memory. When we meet an environment like or similar to an experienced environment, we draw from our long-term memory to relate to the experience. Our working memory is very important as it takes in and retrieves information. But our working memory is lazy and our long-term memory is forgetful. Sometimes our minds work very hard emotionally and reasonably. When that happens we start to believe in our knowing and may start the process that is most important to knowing, practice.

Let me give you two easy examples of this. I am a very good driver. I believe I can drive about any vehicle and do it better than about 90% of people on the road. One of the reasons I am good at driving is that I wanted to do it (emotion), and I had lots of driving experience (reason)—I could drive a tractor at 14 years old. Once I had a driver's license I drove all the time, practice. Thirty years later I had the opportunity to learn to fly an airplane. As good a driver as I am is as bad a pilot as I am. Good pilots fly with me and are as frustrated with my flying as I am with most people's driving. The plane I trained in was a yoke controlled type, that is, it looked like a steering wheel. My brain could not get over this. My brain saw a steering wheel, but a yoke on a taxing airplane is in no way a steering wheel. You can turn the wheel left and right all you want and little to nothing will happen, but push those little pedals where the clutch and brake belong and you will turn left and right. It took me fifty hours to learn to fly an airplane. It takes the average 18-year-old about twenty hours and I am a lot smarter than 18 year olds. So, what made me so dumb? Practice, or what is biologically and psychologically known as synapsis. Car practice trained my brain exactly what to do when presented with a steering wheel, clutch, and brake. My brain literally has grooved pathways to drive and to drive well. I learned this from a repetitive past.

What we learn from the past is not how not to repeat the past but that learning from the past causes us to repeat the past. Simply, we practice it. It is in this way that facts can become our enemy because they are past knowledge.

Remember what we said about myths as guiding us forward, that myths are our future as they guide us forward. I want to repeat that a myth that is a "religio-mathematical" idea is a hard concept that is made even harder when we understand that a myth is that which is not true that creates a truth that is false. I believed any steering wheel configuration was in fact like a car. When I taxied my airplane, this fact was completely false.

But let's apply this to non-physical beliefs. Let me give you a belief system problem. Do you believe in spanking children? Go ahead, answer that question in your head and then defend your answer as well. If you are like the many students I have had write papers on this topic in courses in rhetoric or argumentation, you will answer this question based on how you were raised, that is, spanked or not spanked. You may have all kinds of caveats and situational reasoning as part of your defense, but in the end you believe in spanking or you believe in not spanking children. Your past will be your most salient evidence. Further, I have known and studied people who believe in spanking children who have never raised a hand to a child and people who believe in not spanking children who have been known to give a child a decent swat--so much for belief systems.

Our belief systems inform us, give us talking points, and fail us. If we have not tested them or been tested, we can hold onto them til death do us part. But when we take the steering wheel of our beliefs and turn them left and right and nothing happens, then what? I would say, rejoice! It is time for your brain to cut some new paths to thinking, to knowledge. It is time to grow. I have always liked the Christian Apostle Paul's take on this: When I was a child, I acted as a child, but when I became a man, I put away childish things. We all can grow through challenges. The example I presented from driving to flying is a fairly typical one, that is to say, an easy one. It was knowledge growth, but it was not dynamic knowledge growth. That was not an experience that challenged my thinking but challenged my abilities. Idea knowledge growth comes through the hard work of growing through metaphors that challenge idea realities that have been held too easily against our changing realities. The myths that make our beliefs real and actionable are the myths that move us from slavery to freedom. The Mayan mythmakers tell us this. Let the land, the landscape if you will, tell us this. The ancients who have returned to the land are ever ready to inform us to move forward until we meet the land again. Time will not stand still. The past repeats us. The future actualizes us. Our myths inform us.

What is your myth? I want to open the last discourse with some metaphors that create action rather than beliefs. These are myths that move us rather than mire us in defunct ideology. In the end, I hope you can point to your own myths that give you the very joy of being the most important thing in the world.

## Discourse Four:
## It's There Going

I want to introduce religions as myths for our first look at action myths. And the first religion I want to consider is that of the Tralfamadorians. The Tralfamadorians are fictional aliens who appear in the novels of Kurt Vonnegut. Their religious doctrine is that all time is extant and anyone can visit any piece of that extant time anytime they wish. Every piece of time is always happening. If you want to visit your sixth or sixteenth birthday, go ahead; if you want to kiss your first love or visit your wedding, go ahead; if you want to go to a party in your past or future, be your guest; if you want to go to your birth or death, off you go. As the Tralfamadorians say, you are there, you have always been there, and you always will be there, the moment is simply structured that way.

What Vonnegut has done so ingeniously is take on the illusionists' most magical illusion: time. Time is the illusion of religion and science. Like our birth and death time is a slippery memory. We don't know when time began or when it will end. Still, this does not prevent religion or science from telling us, if not exactly, approximately when time began and when it will end. I would like to tell you that none of them know. And further, it does not matter. And still further I will tell you when time began and when it will end. Actually I can tell you for certain when it began, and if you were dead, I could tell you exactly when it ended. By now you have easily guessed, I am pointing to your birth as the beginning of time and your death as the end of time.

41

We, for certain, know nothing else about time except for recorded history and hopeful future. You are welcome to argue that you have existed before your birth or that you will exist after your birth, but you will be wasting time. So why did Kurt Vonnegut do what he did in creating Tralfamadorians who could bust the chronology of time? Did he really believe we could jump about our time's continuum? No he did not. He simply created a myth that is as interesting and desirable as our life before and life after myths. That is, he wanted us to believe in this myth for reality reasons: to stop being myopic about our lives. Plus he wanted us to embrace myth as a way out of the myopia of bigoted narcissism. An essential quality of the time of our lives always existing is of course a myth that we are here, we always have been here, and we always will be here; that we have some sort of eternal consciousness.

Vonnegut's fictional creation is pretty much like all religious creations of before and after lives we know nothing about. What the Tralfamadorians' interesting little twist does, though, is make our physical existence eternal. We, of course, know this is absurd. But it is instructive in the same way of all religious myths, that is, all religions are instructive. If we had a life before this real one and if we are going to have a life after this real one, what does this tell us about the real one? There are many answers to this, but only one is true and the others are false. Let's consider the Trafalmadorian maxim again: we are here, we always have been here, and we always will be here. This is actually a piece of pure syllogistic logic. Claim one is true, we are here. Claims two and three, we always have been here and we always will be here are dependent on our being here in the first place. Given the evidence of one it is tempting to accept two and three. Here is manifest the truth of myths to be so instructive, but this is too often closely followed by the desire to create fantasies beyond the instructive truths. It is too often these fantasies that become beliefs of a dark and inevitable disastrous reality: duality.

Let's look at how this fits our thesis of 2012's instruction of just how important you are. The time duality of a beginning and an end is the beginning of duality. All religious texts offer a creation and conclusion of time story. In America there are two prophetic religions of the world's end. One is Christianity with its book of Revelations and the other is the Mayan Calendar's end of time. I am not ignorant of the eschatology of Judaism, Hinduism, or Islam. I do not include them because they are typically a mixture of what is characteristic about Christianity or Mayan prefiguration. Also, this is not a discussion about the end of the world, but in fact its apocalyptic beginning. It is about you—the beginning and the end of you--whether you are no religion, some religion, or all about religion. Indeed, comparative religious studies often point to this idea of an end time as a unifying theme. That is to say, they all have it. Because we weren't there in creation and we have not arrived at the end, we can only fantasize about them. They are simply not times we can point to as empirical.

True it is in all religions we have God's word to tell us all about the beginning and the end along with the manifold interpretations of just what God meant. With the Mayan Calendar, we have an ending that is, as we have pointed out

43

in Discourses One and Two, just as full of wild interpretations. These interpretations, like religious ones, depend entirely on some non-real event to become reality, that is, something or someone coming to fix the world of duality and bring some genuinely hoped for eternal peace, which is to say the end of fantastic duality. The Mayan Calendar, on the other hand, does not present some fantastic coming. Remember the Mayan Calendar is a time machine. It marks time not as a matter of prophecy but as a matter of prefiguration—suggesting the future from the past. In the Mayan sense this is not a guess but a reality—a reality not of repetition of the past but of the past representing reality more confidently than prophecy because it is measured in the reality of time. Everyday is a mystery but it is, after all, a day. For the Mayans it could be said, we are here, we were here (the ancients), we will be here. Historically we flourished, contemporarily we are struggling, futuristically, we will be as the sun is and will be. "It is there going."

How does this help us? Honestly, probably no better than the Tralfamadorians, or the Lutherans, or Catholics, or Buddhas, or Baptist, or Mormons, or String Theorists, or evolutionists, or creationists, or New Agers, or Absolutists, or, well you get the idea. I have personally never looked into any of the illusionist's myths that don't have some appealing and appreciable goal of good for humankind. Still, all of these dogmas and formulae never escape some insidious insertion of duality—some reasoned exercise of accusing and even hating a fellow human. The last thing I would ever want to do in present the Mayan Cosmovision as a religion. It is not a religion. Like I said in Discourse One, the vast majority of Mayans are Catholics. If you are looking for a dogma, or pattern, or formula in the Mayan Cosmovision, then you are on the same misguided hunt and gather expedition of so many astropsychologists who are already in print. This does not mean that there is nothing to do. In fact, the Mayan Cosmovision is a most direct path to doing. What does the Mayan Cosmovision indeed tell us to do? It tells us to listen to the land. It tells us to adhere to the reality of the message of the unbroken land of the Americas. It tells us we must end fantasy duality.

When we move away from the illusionist to the scientist view of the Mayan Cosmovision, we find a more substantial methodology for listening to real Mayans and divining what their calendar means. The anthropologists and academic culturists may be sometimes less exciting to read, but so goes the hard work of knowledge. In introducing the more real, if you will, presentation of Mayan predictions, I want to shamelessly paraphrase Charles H. Long, speaking of Robert Redfield (U of Chicago Anthropologist) "the fieldwork has been done. It is now time for Anthropologist to begin to think!" And this should well be true of the work of 2012 diviners: the books have been written, the lectures delivered, and the organizations formed. It is indeed, now time to think! The real question is what to think about. It is time to think about how to make conduct out of divination. To end duality in 2012 if the vibrations don't reverberate, if aliens don't come, if poles don't reverse, if the Americas don't collapse, if capitalism doesn't fold its tent, what will the illusionists do? They will do what they always do: switch, create, or recreate religions fantasy rather than beliefs.

This is the difference between the illusionists and the realists. The realists' belief system is perhaps no stronger than the illusionists, but when reality changes the realists change. The illusionists will never change. The most entrenched will dress their argument in new clothes; that is to say, will change the doctrine of action but not the dogma. In effect, the illusionists will switch religiosity but not religions. If you prove there is no afterlife hell for which there already is no proof, the illusionists will eschew the proof. If you show the reality of hell as a man created reality that too many people live with in life, the right action is a dedicated effort to extinguish the fires of hellish lives. This is the foundational difference between religion and science. Science will discard beliefs in the face of reality. Here is a simple model. When rock'n'roll became the music of America's youth, churches not only attacked its sinfulness but also decreed that people who listened to it were bound for hell. Today, it is a rare church that does not employ rock'n'roll as its church music and as "Christian music." Hell is still around, of course, but the rock'n'rollers aren't bound for it like their pre 1970's brethren. This is a nice social adjustment for religion, but it does not fit the model of changing a belief in light of new evidence or ideas. Instead of considering how powerful religion could be by working to eliminate the hell of reality, it continues to keep hell as an illusion of fear.

The Mayan concept of time is not bogged down in these illusionary practices. 2012 is not an event year for some illusionary prophecy coming to fruition. It is an opportunity year for collective reality—an opportunity to change the time of our life. It is an ancestral prefiguration of our ability to put in time a collective conscious opportunity. It is time to confront duality and use our reality knowledge to end hell. Again, it is more than worth noting here that we are talking about knowledge, not belief. Our illusionary beliefs are the very stuff of hellish invention and violence in the disguise of moral posturing. Yet, I will suggest that 2012 has a morality. What is the 2012 morality? It has nothing to do with creating a moral compass driven by the cafeteria of absolutes that indoctrinate the many human dogma of world religions. For Mayans, aside from having religion forcefully imposed on them by Spanish conquistadors, I suspect this lack of a religious dogma is one of the reasons many Mayans are Catholic. We humans just love to have some proverbial guidance in how to act. Mayan spirituality per se is not about that. It is about being wholly human, physical humans with real lives that have a beginning and an ending. A life that is not designed by God or gods to be anything but a carbon mass with decision making processes. What a glorious and privileged thing to be.

This is proven in Mayan lives by their dedication to Christ's alter ego, Maximon. Maximon is a complicated character who is as iconic as the figures of Jesus. He is the recognition of the reality of life as Jesus is the recognition of the spiritual metaphor of life. Both are characters of sacrifice and salvation. But it is Maximon who recognizes that we are physical. He smokes, drinks, needs money, is lustful, and feels our human pain and diseases. But Maximon does not tell us we exist in error because of our physical maladies. Visiting a Maximon altar is much like visiting a Catholic Church altar: a statuesque icon, flowers, candles, incense, a priest/shaman, prayers and adulation. Maximon will answer prayer. Often there are many testimonies to his graciousness spoken of or even tacked to the walls of a Maximon altar table. But unlike the iconic figures of Catholicism, Mary and baby Jesus, saints, and/or the crucified Jesus; Maximon doesn't want any bead counting chants. He wants tobacco, alcohol, money, blood. If you think this is rather dark, you are mistaken. Maximon is a party guy, full of life and celebration. He is the alter-ego of Jesus without diminishing the true worship of Jesus. He only recognizes that life is real. He celebrates the reality of our physicality instead of denying it. But in this there is no duality of hell and heaven. The duality is erased. The physical and the spiritual are one. It is there going.

And now it is time to eliminate duality. It is time to think about what causes a duality that makes our physiospiritual life one that detracts from our most important status in the world. It is in this way that 2012 is about the beginning work of eradicating violence. Certainly the violence of wars, but not just so obvious a global violence, but the violence of every person in any way to every other person. At my own university a professor was attacked and violently beaten to near death from the "safe" confines of our own university parking garage. This assault was just to take her car. Violence indeed. Have you ever had your car broken into, wallet stolen, lost money in "safe" financial institutions? Have you thought of other people as less than your people? Has the religion you belong to been an historical or current participator in violence against others? Have you been insulted by the accusations of moralists, conventionalists, rule makers or laws which concern themselves with law in the place of justice, mercy, and understanding? Haven't you, the most important person in the world, been a victim of yourself? Only a collective change of consciousness can eradicate these injustices.

Let's begin with a metaphor of war and its obvious violence. In the wonderful film about the Pacific Action in World War II, *The Thin Red Line*, we can find conscious changing moments. After observing and participating in the horrors of battle, the film's narrator, Private Witt, concludes the following:

> War don't ennoble men. It turns them into dogs... poisons the soul.
>
> This great evil. Where does it come from? How'd it steal into the world? What seed, what root did it grow from? Who's doin' this? Who's killin' us? Robbing us of life and light. Mockin' us with the sight of what we might've known. Does our ruin benefit the earth? Does it help the grass to grow, the sun to shine? Is this darkness in you, too? Have you passed to this night?

Private Witt's thoughts mirror a long line of what we call a Natural theme in literature, and this theme is popular in war novels. In fact, it seems only natural that when we see violence on a beautiful day, we are bemused by the violator's inattention to nature. Private Witt feels that staying in the darkness is a choice, and that nature nurtures us toward delight and light. The earth speaks to us. He wonders if you have chosen the darkness and your belief system keeps you there having passed into the night. The Shamans tell us that 2012 implores us to listen to the earth from where we were born and where we will return. Can we finally listen to real time? Can we listen to the time to choose ourselves away from violence, from hell? Can we deal with the reality of time to see the most important thing in the world? A time to breathe in and breath out the mantra of you as the most important thing in the world. Again, think about breathing in as taking care of yourself, and breathing out as taking care of the world. We cannot wait for evolution, religion, vibrations, aliens, governments, or some fantastic comings to transform our consciousness. We are almost there. A time when you are, alas, not only the first person you but the second person you. A time when it is as it has always been: You there going.

# References

These references exist not only to give credit to authors and works cited in the discourses, but to offer the reader a list of works mentioned that did not require citations as well as notes of special interest. I have not included any web sites used in text as they are citied after the notations and web sites have a habit of disappearing overnight.

**Shaman men and women**: The word Shaman itself is a point of argumentation in some academic and indigenous social circles. I recognize the argument, but the over thirty Shamans with whom I had contact were comfortable with appropriating the term. My conversations and ceremonial associations with Kaqchikel speaking Mayan Shamans are at the heart of these discourses. Although I have noted a few, I must respect the wishes of the others to be rather private. I would happily offer them to any serious researcher of Mayan spirituality or way of life, but do not wish to put them out there for pedestrian scrutiny.

**Artemio Hernández, Tojil**: Tojil is a leader among Mayan Shamans. He is an intellectual who interprets dreams and manages a very public Maximon alter at Antigua's Nim Po't. His clarity on Mayan issues was invaluable to me.

Bell, Elizabeth R. Elizabeth is my wife and a PhD Candidate at The Ohio State University and the author of this work's preface. She is currently writing her dissertation on Kaqchikel-Maya spirituality, after multiple trips to Guatemala, culminating in a nine-month fieldwork period in Guatemala through the support of Fulbright.

Brown, Charles Brockden. *Wieland and Memories of Carwin the Biloquist.* Penguin Classics, 1991.

Calleman, Carl Johan. *The Mayan Calendar and the Transformation of Consciousness.* Bear and Co., 2004.

Coe, Michael. *Breaking the Maya Code.* Thames and Hudson, 1999.

*El día y el destino: Desde los derechos hasta el 2012* (Day and destiny: From Rights to 2012). Symposium of Mayan Shamans, Catholic Priests, and Evangelical Ministers: Antigua, Guatemala, August 8, 2011.

Forster, E. M. *Aspects of the Novel.* Harcourt Inc. 1955.

**Gonzalo Ticun, Aq'ab'al**: Aq'ab'al is a model of the indigenous people of Guatemala, but more importantly he is my dear friend. He deserves much credit for offering his sensible, intellectual, and spiritual views of indigenous life in Guatemala. He lives in Santiago Sacatepéquez and is employed as a Kaqchikel language teacher and salesperson at Antigua's Nim Po't.

Leon-Portilla, Miguel. *Time and Reality in the Thought of the Maya.* University of Oklahoma Press, 2nd ed. English Translation, 1988.

Long, Charles H. Introduction in *Rituals of Sacrifice* by Vincent Stanzione, First University of New Mexico Press, 2000.

**Marco Antonio Guaján Cristal, Mokchewan**: A Shaman of great respect and high office in the Kaqchikel Mayan community. Mokchewan was instrumental in my wife and my gaining entrance into the wonderful and mysterious community of Mayan Shamans.

Melville, Herman. *Moby Dick*: Chapter, "The Monkey Rope." Bantam Classic, 1851.

Murrow, Edward R. CBS news cast *See It Now*, March 9, 1954.

*The Thin Red Line*. 20th Century Fox, 1998.

Vonnegut, Kurt Jr. *Slaughter-House Five*. Dell publishing, 1969.

# ABOUT THE AUTHOR

Terrence Bell has worked as an affiliate faculty instructor for Indiana Wesleyan University and an adjunct professor of education and English and assistant men's soccer coach at Capital University in Columbus, Ohio. He has authored articles on education and teacher evaluation. His web and blog sites can be found at twelthsign.com.

## Special Thanks

To Terry Spires not only for his editing and technical help but for his constant *spirituous*

and

To Elizabeth who speaks the One Heart in three languages and to me everyday

and

To my daughter, Madison, who made Guatemala her Spanish learning home and helped me with numerous translations.

www.ingramcontent.com/pod-product-compliance
Lightning Source LLC
Chambersburg PA
CBHW060146050426
42448CB00010B/2332